Danger : inhalants

Chier, Ruth   Test#: 15004

Points: 0.5   Lvl: 3.4

W9-DHJ-692

Lincoln School Library

The Drug Awareness Library™

Danger:
INHALANTS

Ruth Chier

The Rosen Publishing Group's
PowerKids Press™
New York

Published in 1996 by The Rosen Publishing Group, Inc.
29 East 21st Street, New York, NY 10010

First Edition

Book design: Erin McKenna

Photo credits: Cover by Michael Brandt; all other photos by Maria Moreno.

Chier, Ruth.
    Danger: inhalants / Ruth Chier
       p.   cm. — (The Drug awareness library)
    Includes index.
    Summary: Discusses the dangers posed by abuse of chemical inhalants, such as glue, nail polish, and gasoline.
    ISBN: 0-8239-2340-1
    1. Solvent abuse—Juvenile literature. 2. Solvents—Health aspects—Juvenile literature. 3. Substance abuse—Juvenile literature. [1. Solvent abuse. 2. Drug abuse.] I. Title. II. Series.
HV5822.S65C45 1996
363.29'9—dc20   96-14645
                                             CIP
                                             AC

Contents

Different Kinds of Inhalants

Have you ever noticed the smell of the glue you were using for a project? How about the nail polish remover your big sister uses? Or the markers you draw with? Or the gas that your mom or dad pumps into the car? Some of these things may smell good to you. Others may not. One thing they all have in common is that if you breathe them in for too long they can hurt you.

◀ Even markers that smell like fruit or bubble gum can hurt you if you breathe them in for too long.

5

What Are Inhalants?

An **inhalant** (in-HAYL-int) is a **chemical** (KEM-ih-kul) that a person can breathe in. The chemical changes the way that person thinks or feels for a little while. Inhalants are not made for breathing in. They are chemicals like nail polish remover that are used for something else. But some kids breathe them in on purpose. Those chemicals, used as inhalants, hurt kids.

Nail polish remover can hurt you if you use it the wrong way. ▶

What Do Inhalants Do?

Our bodies need air to breathe. When you breathe in, the air goes from your lungs to your brain and all over your body. When someone breathes in an inhalant, it goes into his lungs and brain too. When a person uses an inhalant, he is putting a dangerous drug into his body.

◄ Breathing in an inhalant makes your body stop working the way it's supposed to.

How Do Inhalants Hurt?

Many bad things can happen when someone uses an inhalant. He may become violent and hurt himself or other people. He may fall **unconscious** (un-KON-shus). He may even die.

When a person uses inhalants for a long time, he hurts his body. He hurts his brain, liver, kidneys, throat, nose, and lungs. Most of this **damage** (DAM-ej) can't be fixed. A person who uses inhalants long enough may no longer be able to walk, talk, or think.

Using an inhalant could make you pass out. ▶

Who Uses Inhalants?

Many kids use inhalants. They may learn how from a friend or an older brother or sister.

Someone may have told them that using inhalants is fun. They may think that it doesn't hurt them. But it does. Every time a person uses an inhalant, she damages her body. Since you only have one body, it is important to take care of it.

◀ Sometimes people whom kids look up to try to teach kids to do things that are bad, such as how to use inhalants.

How Inhalants Are Used

People breathe inhalants in three different ways: sniffing, huffing, or bagging. Sniffing means breathing in through the nose. Huffing means breathing in through the mouth. Bagging means breathing in from a bag.

No matter how someone uses inhalants, they will hurt him. There is no safe way to breathe in inhalants.

It is scary to watch someone use an inhalant because you know she is hurting herself. ▶

You Can Tell

You can usually tell when someone uses inhalants. If she sniffs she may have a red, runny nose or her eyes may be watery. If she huffs she may have sores around her mouth. Her breath may smell like chemicals. She may feel sick to her stomach. She may be nervous, moody, or restless. She may have headaches. These are all signs that a person may be using inhalants.

◀ You can't tell how someone will react when she uses an inhalant.

Why Do People Use Them?

Some people use inhalants because their friends or brothers or sisters use them. They want to be cool and fit in. Others use inhalants because they like the way it makes them feel. They may like losing **control** (kon-TROLL) of their bodies. They may like feeling dizzy or woozy. They may want to forget about their problems. Or they may like being able to talk to new people more easily. But each time they use an inhalant, they hurt their bodies.

18

Some people use inhalants to make it easier for them to talk to other people. ▶

You or Someone You Know

You don't want to start using inhalants. If you have tried an inhalant, be smart and don't do it again. If you use inhalants often, tell your teacher, parent, counselor, minister, or rabbi. That person can help you stop using them.

If you know someone who uses inhalants, tell an adult about it. Every time that person sniffs, huffs, or bags he risks his life. You could help save his life.

◀ If you or someone you know uses inhalants, tell an adult you trust. She will be able to help.

21

Saying No

You may have friends who use inhalants. They may ask you to try them too. They may say that you're not cool if you don't try them. Or they may say that you're scared.

But no one can make you do something you don't want to do. And a real friend won't ask you to do it in the first place. If someone asks you to try an inhalant, it's okay to say, "No, thanks. I don't do drugs."

Glossary

chemical (KEM-ih-kul) Basic building block from which things are made.

control (kon-TROLL) Have power over.

damage (DAM-ej) Hurt.

inhalant (in-HAYL-int) Chemical that is breathed in.

unconscious (un-KON-shus) Being unaware of your surroundings.

Index